Make It, Build It!

by Alice Boynton

Red Chair Press   Egremont, Massachusetts

Look! Books are produced and published by Red Chair Press:

Red Chair Press LLC   PO Box 333   South Egremont, MA 01258-0333

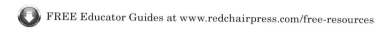 FREE Educator Guides at www.redchairpress.com/free-resources

Publisher's Cataloging-In-Publication Data

Names: Boynton, Alice Benjamin, author.

Title: Make it, build it! / by Alice Boynton.

Description: Egremont, Massachusetts : Red Chair Press, [2021] | Series: Look! books : What a job | Interest age level: 005-008. | Includes index and resources for additional reading. | Summary: "In this book, readers learn about some of the people who use STEM and STEAM skills to make and build things such as engineers and architects"--Provided by publisher.

Identifiers: ISBN 9781634408295 (library hardcover) | ISBN 9781634408332 (paperback) | ISBN 9781634408370 (ebook)

Subjects: LCSH: Engineers--Vocational guidance--Juvenile literature. | Architects--Vocational guidance--Juvenile literature. | Construction industry--Vocational guidance--Juvenile literature. | CYAC: Engineers--Vocational guidance. | Architects--Vocational guidance. | Construction industry--Vocational guidance.

Classification: LCC TA157 .B69 2021 (print) | LCC TA157 (ebook) | DDC 620.0023 [E]--dc23

Photo credits: Shutterstock except for the following: p. 1: iStock; p. 5: dpa picture alliance archive/Alamy; p. 8: Stefan Wackerhagen/Alamy; p. 10: wanderluster/Alamy; p. 12: Tim Larsen/AP Images; p. 15: Richard Drew/AP Images

Printed in United States of America

0920 1P CGS21

Table of Contents

It takes a lot of skills, like math and science, to make and build things. Which of these would you like to see?

Roller Coaster

This job has lots of ups and downs. How high up will the roller coaster go? How fast will it go down? How many times will it twist and turn? **Designers** use science, engineering, and math to figure things out.

Good to Know

Designers want to make the roller coaster fun. But it must be safe, too.

4

A roller coaster has many parts. They are made in a factory. Trucks bring the parts to the site of the ride. Workers put them together there. But no one can ride yet. Is the roller coaster safe? Designers use technology to test it with robots and dummies first.

Roller coasters are tested many times before you ever ride them.

Ice Hotel

Brrrr! Ice hotel builders work where it's cold. First, they put up big metal frames. Then, they spray snow on the frames. Lots of snow! The snow freezes. The builders take the frames away. What's left? Ice walls and roofs.

This Ice hotel is in Sweden. But others can be found in Canada, Finland, and Norway.

9

Workers are busy inside the hotel. *Bzzzz! Whirrr!* They use saws to cut big blocks of ice. They build rooms and furniture, too. Guests will sit on ice chairs. They will eat at ice tables. Even the beds are ice. Would you like to sleep in an ice hotel?

Good to Know

Guests use warm sleeping bags and blankets on their ice beds.

Parade Balloons

Here comes the Thanksgiving Day Parade! The balloons are BIG! It's a big job to make them, too. An artist draws the character. Another artist makes a **model**. What size will the real balloon be? A computer scientist figures it out.

Good to Know

Each part inside the balloon is separate—like the rooms of a house. If an arm or a leg tears, the rest of the balloon stays up in the air.

13

Next, workers cut and sew huge pieces of fabric. The balloon is **inflated.** Are there any leaks? The team waits six hours. The balloon stays inflated. *Whew!* Artists paint it. The balloon is ready to go.

Good to Know

The gas is let out of the balloon. It goes to the parade by truck. Then it is inflated again.

Recycled Houses

The Bottle House Village is built with **recycled** glass bottles. A man and his daughter collected more than 25 thousand (25,000) bottles. They cleaned each one by hand.

The Bottle House Village is on Prince Edward Island, Canada.

The builder used **cement** to glue the bottles together. The houses were popular. Now there is a tavern and a chapel. The chapel was built with 10,000 glass bottles!

Good to Know

A village in Panama is being built with recycled *plastic* bottles.

Corn Palace

The Corn Palace is **unique**. Guess what it's decorated with. Corn cobs! They form pictures. Different colors of corn are used. Red, brown, black, blue, yellow, orange, and even green corn.

The Corn Palace is in South Dakota. Farmers in the state grow lots of corn.

HELL CORN PALACE

2007 ★

21

The cobs of corn are nailed
to the building, one by one.
Visitors love the art work.
So do the birds!

Words to Know

cement: a hard glue-like material to hold things together

designers: people who think of things that can be made or built

inflated: filled with air or another gas

model: a small size of something that is big

recycled: making something new from a thing that was used before

unique: the only one; not found anywhere else

Learn More at the Library

Check out these books to learn more.

Sweet, Melissa. *Balloons over Broadway*. Houghton Mifflin Harcourt, 2011.

Slaymaker, Melissa Eskridge. *Bottle Houses*. Henry Holt and Company, 2004.

Prokos, Anna. *Matsumura's Ice Sculpture*. Celebration Press, Pearson Learning, 2005.

Index

About the Author

Alice Boynton is a writer. She works with others to make books like this one. She has never slept on an ice bed but would like to visit an Ice Hotel.